Dedicated to Asher, Emmy, Evelyn, Grace, Lily, Hunter, and Owen.

The Teeny Tiny Telescope soared through the brilliant, deep, dark sky.

Copyright 2024 by Deanne Dietz

All rights reserved. No part of this publication may be reproduced, stored or transmitted in any form or by any means, electronic, mechanical, photocopying, recording, scanning, or otherwise without written permission from publisher. It is illegal to copy this book, post it to a website, or distribute it by any other means without permission.

Physical Drawings Converted to Digital Drawings by: Moch. Fajar Shobaru from Fiverr - @mfShobaru

To Connect with Author Deanne Dietz:
Website: www.authordeannedietz.com
Instagram: deanne.dietz.author
TikTok: deanne.dietz.author

Saturn chuckled as they saw the Teeny Tiny Telescope flying in a flurry.

As Saturn nodded in approval.

"I'm also made up of the gases hydrogen and helium."

The Teeny Tiny Telescope chuckled with delight and bid adieu from their new friends side.

www.ingramcontent.com/pod-product-compliance
Lightning Source LLC
Chambersburg PA
CBHW040349220526
45473CB00009B/2828